Where Did it Go?

BOOK 1

When sugar disappears and lemons hide in plain sight, AJ and Yoyo discover that not everything that vanishes is gone. Through stirring, tasting, and waiting, they uncover the quiet magic of mixing, melting, and change.

Science inside

Dissolving • Mixing • Solutions

AF811124

Thinking skills

Observation • Wondering • Making Connections

HEERAL SHARMA

Part of the Wonder Loop™ Series
Learning through living, wondering, and doing together.

Published by Happy Orange
myHappyOrange.com
PO Box 266, Tahoe Vista, California

ISBN: 978-1-972521-00-7 (Paperback)

Written by **Heeral Sharma**
Book design by **Purple Parlo**

Illustrations created using **Scrively**
under the direction of the author.

Foreword

Some children learn best by listening.

Some children learn best by watching.

And some children learn best by touching, mixing, smelling, wondering, and discovering.

The Wonder Loop™ Series is created for children who need to experience learning — not just hear about it.

In these stories, science unfolds slowly. There is no rush to find answers. No long explanations. Just shared moments, small questions, and the joy of noticing.

Curiosity grows best when it feels safe.
And sometimes, the most powerful learning begins with

"Hmm… I wonder."

A Wonder Loop™ Book

Read slowly. Pause. Notice. Wonder together.

LOOK ▸ WONDER ▸ ACT ▸ REFLECT

Wonder loop™

Yoyo lay still.
Too still.

His head felt buzzy.
Not loud. Not noisy.
Just full.

"You look like you have many thoughts today."
AJ leaned in.
"They won't stand still," mumbled Yoyo.

"Hmm... let's give them jobs,"
said AJ.

"Do your thoughts like lemonade?"
asked AJ.
Yoyo liked lemonade.

"Sugar?" asked AJ.
"Let's start with 6 spoons."
AJ slid the sugar bowl towards Yoyo.

Yoyo added the sugar.
1... 2... 3... 4... 5... 6...
It sank to the bottom.

Swirl! Swirl!

“It disappeared!” Yoyo exclaimed.

"Try some?"
AJ offered.

"Sweet!
So... it didn't disappear... it's just hiding!"
Yoyo exclaimed.

Squeeze… drip… drip!
Fresh. Tangy. Lemony.
"The lemon is hiding too." Yoyo said.

Sour!
"perhaps, a little more sugar?"
asked AJ.

"Lots of sugar!"
added Yoyo.

"And a pinch of salt... to make it brighter."

AJ smiled.

"Mmm... now it's perfect!"
declared Yoyo.

"Not without the magic of mint!"

AJ rubbed the mint gently.

Cool. Crisp. Minty.

Swirl! Swirl!

Mint stayed... it did not hide.

Clink! Clink!

"And some ice cubes..." added AJ.

Cold. Slippery. Wet.

The ice cubes melted

sl...o...w...ly.

Drip... drip...

A tiny puddle formed.

The pitcher felt cold.
Icy cold.

Cold. Sweet. Lemony.
"I can taste it... even if I don't see it."
Yoyo smiled.

Yoyo's thoughts were happy.
He hugged AJ tightly.

Our world is full of wonder.
Some things hide. Some things stay.
Some things take their time.

And that's okay.

"Are you ready to give it a try?"
Follow the recipe on the next page.

AJ'S Simple Lemonade

Wonder loop™

Wonder Moment

Where did the sugar go?
What happened to the lemon juice?
Why did the ice melt?
What else could you try adding?

You will need:

1 lemon
2 cups water
6 teaspoons sugar (or more)
A pinch of salt
A few mint leaves (optional)
Strawberries (optional)
Ice cubes (optional)
A pitcher
A spoon for mixing

Steps:

1. Add 2 cups of water to the pitcher.
2. Add 6 teaspoons of sugar.
3. Add the juice of one lemon.
4. Add a pinch of salt. Swirl! Swirl!
5. Taste. Too sweet? Too lemony? Add more sugar or lemon juice as needed.
6. Drop in the ice cubes, a few mint leaves, or strawberries (optional).

Smell... Taste... Share... Enjoy!

More Things to Wonder About

You can extend the experience in your own kitchen by **adding different things to liquids** and observing what happens.

Try:

1. Sugar cubes in water
2. Chocolate in warm milk
3. A drop of paint in water
4. Oil in water
5. Salt in water
6. Honey in warm water
7. Flour in water
8. Cornstarch in water
9. Milk in water
10. Oil in vinegar
11. Juice in water

Ask:

Do they stay?
Do they hide?
Do they mix?
Do they float?
What happens when you stir them?
Do they mix the same way?
Does sugar dissolve faster in warm water?
What happens to chocolate in cold milk?

LOOK ▸ WONDER ▸ ACT ▸ REFLECT

Wonder loop™

Wonder Moment

What?
Where?
When?
How?
Why?

Your kitchen is full of discoveries.

The Science inSide

If your child is not yet asking,
"Why?",
you may want to wait before reading ahead.

Re-read this book together several times.
Lean into the surprise:
"What? No way! Where did it go?"

When the world feels a little magical,
children begin to wonder why things happen.

Curiosity comes first before understanding.

When children start asking questions, they are ready
for the science behind what they observed.

Where did the sugar go?

The sugar did not disappear. When you stir it in water, the sugar breaks into tiny pieces. The pieces spread out through the water. They are too small to see, but you can still taste the sweetness. This is called **Dissolving**.

Where did the lemon juice go?

Lemon juice mixes easily with water. It spreads everywhere in the pitcher. That's why every sip tastes sour. When something spreads evenly through water like this, it makes a **Solution.**

Why didn't the mint disappear?

Mint leaves are much bigger pieces.
They do not break apart and spread into the water.
They may add flavor, but you can still see the leaves floating.

What happened to the ice cubes?

Ice is frozen water. When you put ice in lemonade, the warm lemonade shares its heat with the ice, making it melt and become water once again.

Why is there water outside the pitcher?

The cold pitcher cools the air around it.
The air holds tiny bits of invisible water called **water vapor.**
When this warm air touches the cold pitcher, the vapor cools down and turns into tiny drops of liquid water. Those drops gather on the outside of the pitcher.

That's why the pitcher feels cold and wet, even though the lemonade is still safe inside.

Big idea!

Things don't disappear.

They dissolve, melt, mix, change state, or stay the same.

If you look closely, you can discover where they went.

A Wonder Loop™ Book

The Wonder Loop™ is a **simple thinking framework** for finding steady ground when the world feels too big or too fast.

1. Pause II
Slow down.
Take a deep breath.

2. Look >
What do you notice?
What do you see, hear, smell, taste, or feel?

3. Wonder >
What do you think is happening?
Ask: Who? Where? When? How? Why?

4. Act >
What are my choices?
What can I do next?

5. Reflect >
What happened? What did I learn?
What can I try next?

Rooted in RDI® principles of slowing down, using minimal language, and creating space for thinking, it gently guides children to engage in each activity mindfully, building strong episodic memories of competence. It nurtures curiosity, encourages exploration, and supports thoughtful choices. Over time, these small moments of noticing and wondering grow into confident thinking and deeper understanding.

When you are done reading, encourage your child to keep wondering, keep trying, and make a little magic of their own.

Made in the USA
Monee, IL
16 April 2026

48473545R00026